Rock Identification Field Guide

By Patrick Nurre

Rock Identification Field Guide
2nd Edition
Published by Northwest Treasures
Bothell, Washington
425-488-6848
NorthwestRockAndFossil.com
northwestexpedition@msn.com
Copyright 2013 by Patrick Nurre.
All rights reserved. Printed in the United States of America. No part of this book may be reproduced in any manner whatsoever without written permission except in the case of brief quotations embodied in critical articles and reviews.

Rock Identification Field Guide

Introduction	3
The Rock-Forming Minerals	8
The Feldspars	12
Igneous/Plutonic Rocks	14
Igneous/Volcanic Rocks	22
Metamorphic Rocks	37
Sedimentary Rocks	51
Special Rocks	72
Collecting and Building Your Rock Collection	77

Note: Your field guide is color-coded to match the rock kits from Northwest Treasures. This will make match-up of specimens and future ordering easy for you.
- Plutonic Rocks
- Volcanic Rocks
- Metamorphic Rocks
- Sedimentary Rocks
- Minerals

Introduction

Rock identification can be a lot of fun. But it can also hold a lot of frustration. The typical field guide books usually contain pictures that are unrealistic – specimens only found in museums. This field guide book will be an attempt to help you quickly and easily identify and collect the various rock types and their varieties.

First Things First! A good place to begin is at the beginning. Rocks are made of an arrangement of minerals. Minerals are made of a specific arrangement of elements. We cannot see most of the elements with the naked eye – things like oxygen or hydrogen. But when they are arranged into specific and special ways, they form minerals – the stuff of rocks.

And They Were Formed How? So that brings us to the question – how were (or are) elements, minerals and rocks formed? That is a great question. No one was around to see the first elements, minerals and rocks form. No one observes their formation today, with the exception of volcanic rocks. Many geologists think they know. But they were not around at the beginning either. They make guesses based on human reasoning called uniformitarianism – *the present is the key to the past.* This was an idea that originated in the late 1700s during a time of human questioning about the existence and nature of God. They reasoned that the Scriptures (the Holy Bible) were not reliable. So, they jettisoned the ideas expressed in the Scriptures about the 6 day creation and global flood and came up with their own ideas. The early secular geologists reasoned that processes they observe today are sufficient to explain the history of the Earth. And since these men had already rejected the idea of a young creation and global flood, it seemed a simple task to imagine the same forces that are active today sculpting the Earth over very long periods of time. Of course they were way off the mark. The creation and global flood as taught in the Scriptures were, up to the time of the Enlightenment, considered to be real historical events, not myths or nice stories. If the Flood of Genesis really did happen, then it would have undoubtedly left its mark in the rocks all around us.

What secular geologists concluded was that the Earth was a giant recycling set of geologic processes, not really having a beginning and not really having an end, but constantly changing by recycling Earth materials over and over. They say that what we observe today is simply the result of that ancient recycling process. And so the rock types are studied and categorized according to this idea.

Rock Types In secular geology there are 3 major rock types all based around their supposed origin. The major rock categories we have today in geology are the result of this reasoning. Here is the basic approach to the modern rock type categorization:

1. **Igneous Rocks** – those rocks formed by fire, either by volcanic eruption (lava, "extrusive" rocks) or from magma deep under ground. Another name for these types of rocks are "intrusive" rocks because it is believed that they formed deep underground by hot magma, cooled over millions of years and then *intruded* into the Earth's crust, forming great granitic formations.

2. Metamorphic Rocks – those rocks formed by heat and pressure deep underground, changed from preexisting rocks over millions of years.

3. Sedimentary Rocks – those rocks they think formed from advancing seas as they advanced and retreated over hundreds of millions of years across the Earth's landscape, leaving in some cases a record of plant and animal life. Secular geologists believe that these rocks record the story of the evolution of life from simple to complex. In fact sedimentary rocks are dated according to the **belief** that life developed from simple to complex: the simple life being in the lowest rock layers, followed by increasingly more complex life like fishes, amphibians, reptiles, mammals and finally man. Sedimentary rocks indeed record a history, but it is a history of catastrophe and extinction, not of evolution.

A Different Approach Now that we have briefly looked at how secular geologists group the rocks, let's take a look at the Biblical approach. We will do what Jesus taught in Matthew 4:4 where He said, *"Man shall not live on bread alone, but on every word that proceeds out of the mouth of God."* This would include the 6 day creation of Genesis chapter 1 and the global flood of Genesis chapters 6-9. This is exactly how the prophets and the apostles reasoned. In 2 Peter 3:3-10 we read, *"Know this first of all, that in the last days mockers will come with their mocking, following after their own lusts, and saying, 'Where is the promise of His coming? For ever since the fathers fell asleep, all continues just as it was from the beginning of creation.' For when they maintain this, it escapes their notice that by the word of God the heavens existed long ago and the earth was formed out of water and by water, through which the world at that time was destroyed, being flooded with water. But by His word the present heavens and earth are being reserved for fire, kept for the Day of Judgment and destruction of ungodly men. But do not let this one fact escape your notice, beloved, that with the Lord one day is like a thousand years, and a thousand years like one day. The Lord is not slow about His promise, as some count slowness, but is patient toward you, not wishing for any to perish but for all to come to repentance. But the day of the Lord will come like a thief, in which the heavens will pass away with a roar and the elements will be destroyed with intense heat, and the earth and its works will be burned up."* This brief passage of Scripture touches on the geology of the Earth!

- The Earth was formed out of water and by water
- The Earth of that time was flooded with water
- The present space, Earth and everything in them will be burned up
- The elements will melt with intense heat
- There will be a new heavens and new earth

Uniformitarian (secular) geology teaches that the Earth began about 4.6 billion years ago out of melted rocks and lava. In fact one of the reasons secular geologists conclude this is that it somewhat explains why we have so much volcanic rock on our Earth today. But could there be another legitimate explanation for this? *If we include the Genesis Flood, yes, there is.* It appears that the real fire will take place later but it was not a part of the creation of the Earth!

For purposes of this guide, the term "secular geology" refers to the study of geology without reference to the Bible.

As always, if you run into a question or problem, you can always drop me an email or give me a call. I am always glad to hob-knob with a fellow rockhound.

So, let's start looking at the rocks in a different way.

Patrick Nurre
August 2013
northwestexpedition@msn.com
425-488-6848

The Rock-forming Minerals

Although there are over 4,000 identified minerals, in rock identification fortunately you only have to know a few to identify most of the rocks. The rock-forming minerals are grouped into dark and light colored minerals. Knowing these minerals will help you identify most of the rocks you will discover. Below are pictures of the most common rock-forming minerals.

Examples of the dark colored rock-forming minerals:

Pyroxene (augite)

Amphibole (hornblende)

Iron (magnetite)

Calcium feldspar (labradorite)

Biotite (black mica)

Olivine

Examples of the light colored rock-forming minerals:

Quartz

Jasper (opaque quartz, and occurs in a variety of colors due to iron)

Potassium feldspar

Sodium feldspar (although it is the primary, light-colored feldspar in the darker volcanic rocks)

Muscovite (white mica)

Calcite (primarily makes up
sedimentary rocks like limestone)

Review these rock-forming minerals and get thoroughly acquainted with them. They hold most of the key to identifying the various rock types.

The Feldspars

As you can see from the list of rock-forming minerals, feldspar plays a huge role. So, I would like to include additional information on these amazing minerals. You will most likely only be concerned with the most common – orthoclase, labradorite and albite, but you will from time to time run into the others too.

The feldspars are divided into two basic groups:

Potassium Feldspars or "K-spars"; (The alkali feldspars)
- Microcline (the green variety of which is called amazonite)
- Orthoclase (typically pink to salmon)
- Sanidine
- Adularia (the moonstones; gray, peach and green)
- Perthite is really a combination of either microcline (green) and albite or orthoclase (pink) and albite in a laminar or layering effect

Plagioclase Feldspars (consisting of the minerals sodium to calcium, albite to anorthite)
- Albite (sodium) – lightest in color
- Oligoclase (sunstone; and rainbow moonstone)
- Andesine
- Labradorite and Larvikite
- Bytownite
- Anorthite (Calcium) – darkest in color

Examples of the Feldspars

Potassium feldspars – microcline, orthoclase, sanidine, adularia and perthite

Plagioclase feldspar – albite, oligoclase, andesine, labradorite, bytownite, and anorthite

Igneous Rocks

The word igneous comes from a Latin word for fire. It is believed by secular geologists that rocks like granite and lava all originated in fire from molten magma. We have seen volcanic rocks form. But rocks like granite, no one has seen form. So, what are they?

Let's come up with a little different division and divide igneous rocks into two groups – **plutonic** and **volcanic**. We will group these according to their rock grain size and not their origin.

Plutonic Rocks

The first group of rocks we will tackle is the **plutonic** rocks. The word plutonic comes from the mythological god of the underworld – Pluto. This will reflect the Biblical origin of these rocks. Even though they are not mentioned in the Bible, because Peter states that the Earth was formed out of water and by water, we can suspect that plutonic rocks were the original created "basement" rocks. These are the ones that formed the foundation of the Earth. Another way to group these rocks is to call them granitic or coarse-grained rocks. By coarse-grained rocks we mean those rocks where the mineral crystals that make up the rock can be easily seen with the naked eye. These rocks are in turn categorized according to the light and dark colored

Minerals they are made of. For now we will consider four common rocks that comprise this group.

1. **Gabbro** – this rock got its name from an area of Italy where the rock is typical. All plutonic rocks are identified by the amount of quartz (a lighter colored mineral) they contain in comparison to the amount of darker minerals they contain. Gabbro is a plutonic rock that contains very little quartz and so it is going to have a darker appearance in the field. Below are some pictures of gabbro. Study these samples so that you can readily learn how to identify gabbro when you are in the field.

Examples of Gabbro

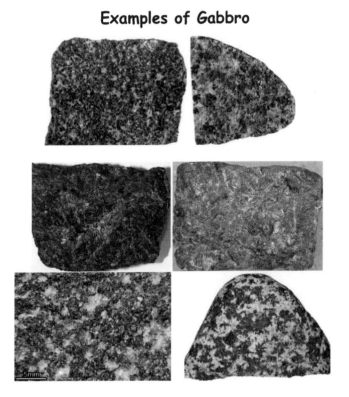

The predominant minerals in gabbro are: pyroxene (augite), plagioclase (both sodium and calcium feldspar), amphibole (hornblende), biotite mica and olivine.

Take a closer look at the gabbro pictured below and describe what you see.

Do you see the predominance of the darker minerals? The lighter colored mineral that you see in the first picture is plagioclase feldspar, but of the lighter color - sodium feldspar. In the second picture, plagioclase feldspar is the darker feldspar - calcium. Geologists call the darker colored rocks, *mafic*, rocks rich in magnesium (*ma*) and iron (*fic* - from the Latin).

The keys in identifying gabbro are:
    a) Coarse grained - the mineral crystals can be seen with the naked eye
    b) The darker minerals are predominant

2. **Granite** - at the other end of the spectrum is a plutonic coarse grained rock called granite. The word "granite" comes from the Latin granum, a grain, in

reference to the coarse-grained structure of such a crystalline rock. It belongs to the *felsic* group of rocks: those rich in feldspar (the lighter colored feldspars, *fel*) and quartz (*sic*) for silicon dioxide. The combination of these minerals makes it a lighter colored rock. It most often is composed of mostly quartz and potassium feldspar with a bit of black mica and or hornblende.

Examples of Granite

The predominant minerals in granite are: quartz, feldspar (white or pinkish), amphibole (hornblende), and or biotite mica. These minerals also belong to the rock-forming mineral group.

Examples of Minerals in Granite

Take a closer look at the granite in the following pictures and describe what you see.

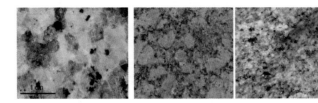

The keys in identifying granite are:
   a) Coarse grained – the minerals can be seen with the naked eye.
   b) The lighter minerals are predominant

Plutonic and volcanic rocks have both dark and light colored rocks that reflect the color of minerals they contain. These two types of rocks also have *intermediate* rocks between the darker colored and the lighter colored rocks. The intermediate rocks are more difficult to identify, especially in the volcanic rocks. The plutonic rocks are not as difficult, but will take some practice in knowing what to look for.

3. Diorite – an intermediate coarse grained plutonic rock between gabbro and granite which has a spotted appearance that has earned the nickname of the "salt and pepper" rock. The salt and pepper appearance is due to an even mix of both the light colored and the dark colored minerals. Diorite also has very little to no quartz in its make-up.

Examples of Diorite

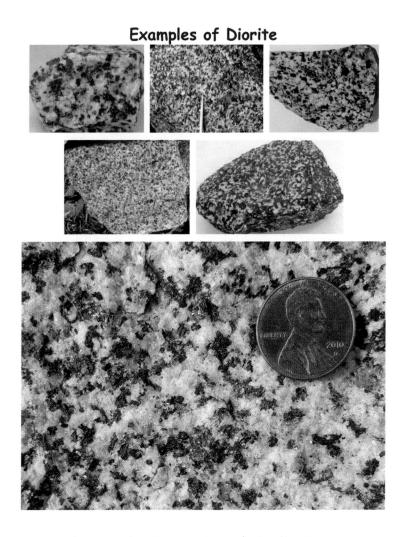

There are a few predominant minerals in diorite – plagioclase feldspar (white), pyroxene, hornblende and biotite.

The keys in identifying diorite are:
a) Coarse grained - the minerals can be seen with the naked eye
b) An even mix of the lighter colored and darker colored minerals (plagioclase feldspar and pyroxene, hornblende and biotite)
c) Almost no or very little quartz

4. Granodiorite - as the name implies, is a coarse grained plutonic rock intermediate between granite and diorite. It is also called a salt and pepper rock and it is easily mistaken for diorite. But the **major difference** is that granodiorite possesses quartz. It also has an abundance of hornblende and biotite mica. If you use your magnifying glass you will notice the quartz in the make-up of granodiorite. Take a look at the photos below.

Examples of Granodiorite

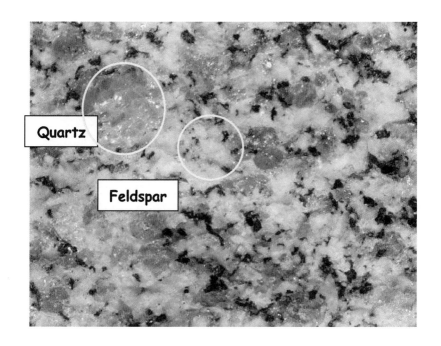

The keys in identifying granodiorite are:
a) Coarse grained – the minerals can be seen with the naked eye
b) Has an abundance of hornblende and biotite mica
c) Has some quartz (translucent mineral)
d) Has an abundance of the lighter colored feldspar

These are the four basic plutonic rocks that you will be most concerned with. There are others that are mixes of these, but they are not easily recognized. If you cannot identify a specific coarse grained rock, the best thing is to identify and record its location and call it a plutonic rock. In time with practice, you may be able to identify some of these "hybrids".

Volcanic Rocks

Also grouped with the igneous rocks by uniformitarian geologists are the **volcanic** rocks. Geologists have observed this group in the process of forming, so we know quite a bit about how volcanic rocks form.

The word volcanic comes from the mythological god of fire – Vulcan. That makes sense. But it is here that the uniformitarian philosophy of the origin of volcanic rocks has erred. Uniformitarian geologists use the present observation of volcanic rocks to explain the volcanic rocks that formed in the past. But Peter tells us in his second letter that the Earth was formed out of water and by water. Therefore from Biblical revelation we understand that the origin of the Earth was not in volcanic eruptions. Where did all the lava on the Earth come from? It is certainly not being produced like that today.

Genesis 7:11 might hold the key to the abundance of volcanoes and lava found on the Earth today. Verse 11 reads, *"In the six hundredth year of Noah's life, in the second month, on the seventeenth day of the month, on the same day all the fountains of the great deep burst open and the floodgates of the sky were opened."* Volcanoes are essentially fissures or breaks in the Earth through which magma flows up and onto the surface of the Earth. Lava is just the name given to this magma or molten rock. Volcanoes seem to be the direct result of the breaking up of the fountains of the great deep. Volcanic rocks are also called "extrusive" rocks because they come out of the ground and onto the surface as lava or other pyroclastic materials. How do I know when I have found a volcanic rock?

Volcanic rocks contain the same rock-forming minerals that plutonic rocks contain except volcanic rocks are characterized by being fine grained. That is, in volcanic rocks one sees a general color, but not the individual mineral crystals that make up the volcanic rock. Geologists call this phenomenon, aphanitic, meaning, "invisible". As with plutonic rocks, there are dark colored as well as light colored rocks and there are also intermediate rocks.

Dark Colored Volcanic Rocks

1. **Basalt** – comes from a Latin word meaning "hard stone". And it is very hard! Basalt contains predominantly the following darker minerals: pyroxene (augite), iron, calcium feldspar (calcium-rich plagioclase), olivine and the lighter colored plagioclase, sodium feldspar.

Examples of Basalt

(Basalt)

 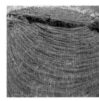 

2. **Vesicular Basalt and Scoria** - Basalt with vesicles or gas pockets is called vesicular basalt. Scoria is a form of vesicular basalt but predominantly filled with many more vesicles than vesicular basalt. It is extremely light weight and has been compared to the lighter colored rock, pumice.

Examples of Vesicular Basalt and Scoria

3. **Oxidized basalt** – basalt which contains an abundance of iron. As a result when it is exposed to oxygen it literally rusts.

Examples of Oxidized Basalt

4. Olivine basalt – basalt also contains a lot of magnesium in the form of olivine or peridot.

Examples of Olivine Basalt

Example of Olivine Basalt with Peridot

Light Colored Volcanic Rock

Rhyolite – is a fine grained lighter colored volcanic rock. It is the predominant lighter colored volcanic rock. Like the lighter colored plutonic rocks, volcanic rocks have a lighter colored version, too. Rhyolite comes from a Greek word meaning "flow". Unlike basalt, rhyolite is extremely viscous. This word has to do with the resistance to flow. What makes rhyolite resistant to flowing or moving quickly? It is the amount of quartz that makes rhyolite sticky and pasty. In fact it often preserves the flow patterns as it cools. The predominant minerals in rhyolite are quartz, potassium feldspar and normally very little iron. Colors range from white to gray to red and everything in between. Because of its quartz content, rhyolite makes great jewelry and polished pieces for display. It often contains spherulites. These are small rounded globules of silica – mini thunder eggs if you will.

Examples of Rhyolite

Examples of Rhyolite with Spherulites

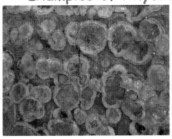

More Examples of Rhyolite

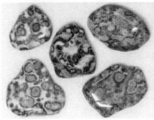
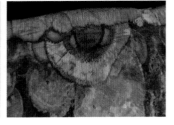

The Intermediate Volcanic Rocks – These can be very difficult to identify because there is such a variety. Generally the intermediate volcanic rocks are shades of gray. The intermediate rocks include:

1. **Andesite** – a fine grained volcanic rock in which the predominant minerals are iron, pyroxene, hornblende and sodium feldspar, the lighter colored feldspar. The sodium feldspar appears as rectangular whitish blocks in andesite.

Examples of Andesite

  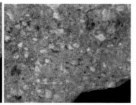

Andesite is a typical lava produced by explosive volcanic eruptions like the stratovolcanoes of the Northwest United States.

2. Dacite – an intermediate lava between andesite and rhyolite. Dacite consists mostly of plagioclase feldspar with biotite, hornblende, and pyroxene (augite and/or enstatite). It has quartz as rounded, corroded phenocrysts (conspicuously larger crystals than the crystals in the ground mass), or as an element of the ground-mass.

The plagioclase ranges from oligoclase to andesine and labradorite. Sanidine occurs, although in small proportions, in some dacites, and when abundant gives rise to rocks that form transitions to the rhyolites. The groundmass of these rocks is composed of plagioclase and quartz. As you can see from the definition of dacite, it is not easy to identify. Like rhyolite it is also very viscous and often preserves flow banding patterns. It is rarer than rhyolite and is often confused with rhyolite. Here are some pictures of the types of feldspars contained in dacite lava.

Examples of the Feldspars in Dacite

Plagioclase Andesine,   The "K-spar" Sanidine,   Plagioclase Calcium

Examples of Dacite

  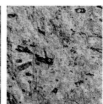

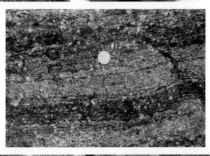

(Dacite)

The best advice for identifying these rocks in the field is if you cannot readily identify the rock, record the location and call it an intermediate volcanic rock.

The Porphyrytic Volcanic Rocks - Although volcanic rocks are fine grained, they often contain phenocrysts in their ground mass. Some of these rocks can be so filled with porphyry that the rock could be confused with a plutonic rock. The phenocrysts are usually of sodium feldspar, pyroxene, potassium feldspar or hornblende.

Basalt Porphyry

Andesite Porphyry

Dacite Porphyry   Rhyolite Porphyry

The Glassy Volcanic Rocks - Rhyolite lava contains a lot of quartz and sometimes this may be produced as obsidian. That's right. Obsidian is volcanic rhyolite glass! Why is it black then? It is most commonly black because of the amount of iron it contains. But there are also examples of "pure" obsidian too.

Examples of Obsidian

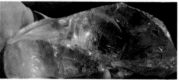

Obsidian sometimes occurs with a white snowflake like mineral called crystobolite, a type of quartz.

Example of Snowflake Obsidian

The darker volcanic rocks can also produce glassy basaltic rocks, although nothing like pure glassy obsidian.

      Basanite   Olivine Basanite   Picrite
(sometimes called Tachyllite)

Apache Tears (rounded nodules of obsidian)

Palagonite - volcanic glass altered by water such that it is extremely fragile and often changes color

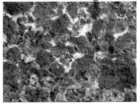

Perlite – a type of volcanic glass altered by water; it is extremely brittle and crumbles easily

The Pyroclastic Volcanic Rocks- The word pyroclastic means, "fire broken". It refers to the tiny shards of glass and volcanic rocks that make up this material.

1. **Ash** – this word is really a misnomer because it conjures up the picture of the ash of burning paper or wood. Volcanic ash is actually small bits and pieces of volcanic rock that moves within a cloud of hot steam up to several hundreds of miles an hour. It acts like a sandblaster as it strips bark off of trees. Some ash is lighter in weight and is carried high up into the atmosphere where it circles the globe and is deposited as the grains become too heavy.

2. **Ashfall tuff** – this material is ash that piles up and because of the retained heat, welds together into a fragile powdery volcanic rock.

3. **Welded tuff** – is very similar to ashfall tuff, but it consists of various sizes of volcanic shards of rock and glass. It welds into a very hard rock.

4. **Ignimbrite** – this term is not used much anymore, but it is essentially welded tuff with larger phenocyrsts present.

5. **Pele's Tears** – hot volcanic glass that is "frozen" in mid air as it is hurled from its original location. It is stretched into all kinds of interesting shapes.

6. **Tephra** – any pyroclastic material.

7. **Bombs** – usually reserved for basalt lava that has been ejected from erupting lava and "freezes" into various shapes and sizes before it hits the ground.

8. **Pumice** – light colored and airy; it is mostly gas pockets and glass.

Ash   Ashfall Tuff

Welded Tuff

Ignimbrite

Pele's Tears   Basalt Bombs

Pumice

Metamorphic Rocks

The word metamorphic is actually a Biblical word. It occurs in Romans Chapter 12:2, *"And do not be conformed to this world, but be **transformed** (changed) by the renewing of your mind, so that you may prove what the will of God is, that which is good and acceptable and perfect."* So the idea behind metamorphic rocks is that they have been changed from their original rock through heat and pressure.

Metamorphic rocks are divided into two categories:

1. Foliated - refers to repetitive layering in metamorphic rocks. Each layer may be as thin as a sheet of paper, or over a meter in thickness. The word comes from the Latin folium, meaning "leaf", and refers to the sheet-like planar structure. Foliation also would refer to the banding in metamorphic rocks where the original minerals are separated out into alternating layers of light and dark colored bands.

2. Nonfoliated - refers to the crystallization of a rock.

Secular geologists guess that it took millions of years and deep burial of preexisting rocks under a great deal of heat and pressure to form metamorphic rocks. But that is conjecture. Vast amounts of rock grating against other rocks can create huge amounts of friction and thus heat and pressure. This could have been very easily produced by the Genesis Flood during the tectonic energy associated with the breaking up of the fountains of the great deep in Genesis 7:11. Realize that once the Genesis Flood is rejected as a possible mechanism for all the geology we see in the earth today, then, anything is indeed possible. But the Genesis Flood has been revealed to us as a real historical event and that would supersede any possible idea advanced by man.

The Foliated Metamorphic Rocks

1. Gneiss – upon careful examination of any sample of gneiss, it becomes apparent that it has the same minerals as the plutonic rocks. It is just banded or layered as opposed to being evenly mixed. It's fair to conclude that most gneiss might have been granite, diorite or gabbro before it was metamorphosed. Examine the following examples of gneiss and try to guess what the original rock was. The minerals are the same in plutonic rocks.

Examples of Gneiss

(Gneiss)

Diagram of minerals in gneiss

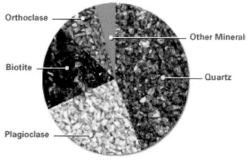

2. **Migmatite** – similar to gneiss is the metarmophic rock, migmatite. The main difference is that it shows much

more metamorphosis in the way of twisting and contorting of the mineral banding.

Examples of Migmatite

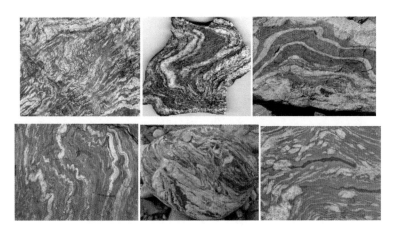

3. **Schist** – metamorphosed layering but with the addition of mica. Perhaps schist was formerly shale or sandstone. No one has ever seen a metamorphic rock form, so we don't really know. But the best way to describe schist is that it looks like layers of glitter. Many examples of schist are very fine grained so that rather than large visible flakes of mica, the texture may appear to be tiny fine crystals cemented together. Schist can also occur in a variety of colors and often can contain large crystals of garnet, staurolite or some other gemstone.

Examples of Schist

(Schist)

4. Slate – appears to be metamorphosed shale, which is a sedimentary rock. Whereas shale is brittle and fragile, slate is foliated, but hard. Slate can occur in all kinds of colors.

Examples of Slate

5. Phyllite – Phyllite appears to be metamorphosed slate with a very shiny appearance as a result of mica.

Examples of Phyllite

The Nonfoliated Metamorphic Rocks-
The second grouping of metamorphic rocks belongs to the nonfoliated family. These rather than being layered or banded, have probably been recrystallized through heat and pressure.

1. Quartzite – as the name implies, is filled with quartz. It is thought to be metamorphosed sandstone. Its grains can be fine or coarse and there is a tremendous amount of variation in the color. It is quite abundant and because it is so hard, polishes beautifully. It is a fun rock to collect because of its variety. And it is easily recognizable, being kind of like a sugar cube in texture. It often preserves the iron-rich bands common in regular sandstone.

Examples of Quartzite

2. Marble – metamorphosed recrystallized limestone; it is rich in calcium carbonate. Marble is quite varied in its patterns and colors. One of the ways to identify marble is by the use of a strong acid. It will react and fizz.

Examples of Marble

3. Amphibolite – Amphibolite is a coarse-grained metamorphic rock that is composed mainly of green, brown or black amphibole minerals and plagioclase feldspar. The amphiboles are usually members of the hornblende group.

Examples of Amphibolite

4. Soapstone (steatite) – Soapstone is a metamorphic rock that is composed primarily of talc, with varying amounts of chlorite, micas, amphiboles, carbonates and other minerals. Because it is composed primarily of talc it is usually very soft. Soapstone is typically gray, bluish, green or brown in color, often variegated. Its name is derived from its "soapy" feel and softness. This rock is not to be confused with the mineral, talc, which makes up soapstone. Without handling the rock, it initially looks like marble or even quartzite. However, it is very soft, easily carved and can be scratched with the fingernail. And it will have a soapy feel to it.

Examples of Soapstone

5. Hornfels – comes from a German word meaning, "hornstone", after its frequent association with glacial "horn peaks" in the Alps. It is a very hard rock and thus more likely to resist glacial action and form horn-shaped peaks such as The Matterhorn. The most common hornfels (the biotite hornfels) are dark-brown to black with a somewhat velvety luster owing to the abundance of small crystals of shining black mica. The lime hornfels are often white, yellow, pale-green, brown and other colors. Green and dark-green are the prevalent tints of the hornfels produced by the alteration of igneous rocks. Although for the most part the constituent grains are too small to be determined by the unaided eye, there are often larger crystals of cordierite, garnet or andalusite scattered through the fine matrix, and these may become very prominent on the weathered faces of the rock.

Examples of Hornfels

Hornfels is not an easy rock to identify. It can be mistaken for basalt or even gneiss. It is an ugly rock, but a necessary one in the understanding of metamorphic rocks. Very similar to it is the next one – claystone or catlinite.

6. Claystone; catlinite; pipestone; pipeclay – is metamorphosed mudstone. Remember that things that have been metamorphosed have been baked, heated and or pressurized so that its basic nature changes. It is a hard, usually reddish brick stone. The term Catlinite came into use after the American painter George Catlin visited the quarries in Minnesota in 1835; but it was Philander Prescott who first wrote about the rock in 1832, noting that evidence indicated that American Indians had been using the quarries since at least as far back as 1637.

Examples of Claystone

7. Serpentinite – describes a great variety of metamorphic rocks containing one of the many serpentine mineral varieties. Geologists think that serpentinite is a metamorphosis of ultramafic rock: rock derived from the mantle of the Earth. So it is naturally going to be a darker colored rock. As you can see from the examples below, it contains a lot of various shades of the green serpentine minerals.

Examples of Serpentinite

The Sedimentary Rocks

Sedimentary rocks are all those rocks that have been laid down by water, mud and wind or are the results of organic deposition. They include loose material like sand and clay and also material that has been cemented together into more or less solid rocks. So long as the material has not been altered from what it was originally when it was laid down, the rocks are considered to be sedimentary.

Secular geologists account for the abundance of sedimentary rocks all over the Earth by way of uniformitarian assumptions – through the course of hundreds of millions of years many and successive shallow seas have encroached onto the continents leaving layer upon layer of sedimentary rocks. The one main agreement between uniformitarians and catastrophic flood geologists is the WATER! In Flood Geology the sedimentary rocks were laid down successively, catastrophically, and in quick succession mostly within the year of the Genesis Flood. As the Flood waters started to recede, the freshly laid sediments were incised and carved into the many landforms we see today. Shortly after the Flood, an ice age set in due to the rapidly changing climatic conditions brought on by many volcanic eruptions and warmer oceans. Most of the jagged mountain peaks visible today were the result of this catastrophic ice age. Toward the end of the ice age catastrophic melting again caused many of the Earth's landforms to be incised and carved like the Scablands National Monument in eastern Washington.

Courtesy USGS.

The Sedimentary Rocks are divided according to the following scheme:

1. Clastic Sedimentary Rocks – the word clastic means "broken". So these rocks are broken bits and pieces of other rocks which have been cemented together. Accordingly clastic sedimentary rocks are grouped according to "clast" size from clay-sized (extremely fine) all the way up to boulder-sized.

2. Chemical Sedimentary Rocks – these rocks are those which have been produced through the process of precipitation from mineral saturated sediments.

3. **Biochemical Sedimentary Rocks** – these rocks have been formed from the remains of plants and animals.

The Clastic Sedimentary Rocks:

1. **Claystone (not to be confused with catlinite, the metamorphic claystone), mudstone and shale.** The clasts which form these rocks are very tiny and fine, usually not seen with the naked eye. These are the most abundant sedimentary rocks on Earth.

Examples of Claystone or Mudstone

(Claystone)

2. Siltstone – from fine grained we move to silt-sized clasts, a little larger than clay-sized and can be seen with the naked eye. Although often mistaken as shale, siltstone lacks the laminations which are typical of shale. Siltstones may contain concretions.

Examples of Siltstone

(Siltstone)

3. **Sandstone** – consists of sand-sized particles of primarily quartz. These clasts can easily be seen with the naked eye. It is the most easily recognizable of the sedimentary rocks. The reddish color in sandstone is due to the iron content which has been oxidized.

Examples of Sandstone

(Sandstone)

4. Arkose – is actually a type of coarse-grained sandstone with the inclusion of potassium feldspar, hence its pinkish color.

Examples of Arkose

(Arkose)

5. **Graywacke, greywacke** or just **wacke** – is from a German word that means a gray, earthy rock. It is sometimes referred to as "dirty sandstone". It is a variety of sandstone generally characterized by its hardness, dark color, and poorly sorted angular grains of quartz, feldspar, and small rock fragments set in a compact, clay-fine matrix.

Examples of Graywacke

6. **Conglomerate** – is a clastic sedimentary rock made up of ROUNDED small to large pebbles, stones, cobbles or even boulders. It is easily recognizable by the rounded stones it contains. These stones can be of any material including the mineral quartz and jasper.

Examples of Conglomerates

7. Breccia – is from an Italian word meaning "breach". Breccia is a rock composed of broken (ANGULAR, not rounded) fragments of minerals or rock cemented together by a fine-grained matrix, that can be either similar to or different from the composition of the fragments. There are many types of breccias and these are designated by the prefix word of its descriptive type.

- **Examples of Sedimentary Breccia** - formed by either submarine debris flows, avalanches, mud flow or mass flow in an aqueous medium.

- **Examples of Volcanic Breccia** – formed from the broken pieces of volcanic rock usually in a welded tuff.

  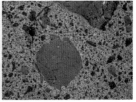

- **Examples of Intrusive (plutonic) breccia** – plutonic clasts within a plutonic matrix.

- **Examples of Metarmorphic Breccia** – metamorphosed clasts within metamorphosed rock. Some marble is metamorphic breccia.

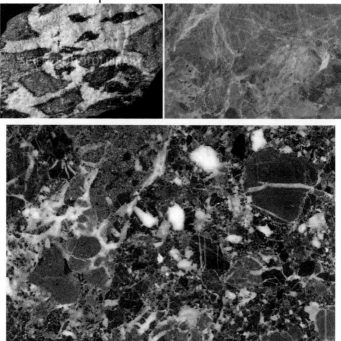

8. Loess - is a clastic, predominantly silt-sized sediment, which is formed by the accumulation of wind-blown dust. This is considered to be largely a product produced by strong winds subsequent to the ice age. In fact many of the "frozen" mammoths in the far north are buried in glacial loess. The word is traced back to the German word, "locker" and then through the English, "loose".

Loess is homogeneous, porous, friable, pale yellow or buff, slightly coherent, typically non-stratified and often calcareous. Loess grains are angular with little polishing or rounding and composed of crystals of quartz, feldspar, mica and other minerals. Loess can be described as a rich, dust-like soil. Loess deposits may become very thick, more than a hundred meters in areas of China and the Midwestern United States. It generally occurs as a blanket deposit that covers areas of hundreds of square kilometres and tens of metres thick.

Examples of Loess

9. Glacial Till – is an unconsolidated mass of boulders, pebbles, sand and fine clay left behind by glaciers when they melted. The boulders and pebbles, can show some wear or they can be tumbled by water and then consolidated as till.

Examples of Glacial Till

10. Tillite – Glacial till that is consolidated into solid rock. It usually consists of a variety of different kinds of rocks.

Examples of Tillite

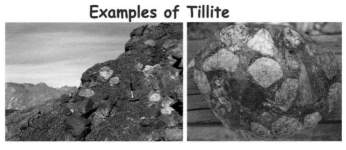

(Tillite)

The Chemical Sedimentary Rocks:
1. **Limestone** – is a sedimentary precipitate rock composed largely of the minerals calcite and aragonite, which are different crystal forms of calcium carbonate ($CaCO_3$). Many limestones are composed from skeletal fragments of marine organisms such as coral or foraminifera (protozoa). Limestone makes up about 10% of the total volume of all sedimentary rocks. Some limestone formations around the world are vast, mountains of sedimentary rock. Of course the Genesis Flood explains this as catastrophic deposition of lime mud sediments some of which is saturated with marine fossils. As limestone can be confused with other sedimentary rocks, a good way to distinquish it is to place a drop of strong acid on it. If it fizzes, it is limestone. The calcium carbonate reacts to acid.

Example of Limestone

Examples of Fossil limestone

Examples of Limestone (without fossils)

Examples of Chalk limestone (fossils of diatoms)

Examples of Coquina (from the spanish for shells) limestone with visible fossil shells

2. Chert – is a fine-grained silica-rich microcrystalline, cryptocrystalline or microfibrous sedimentary rock that may contain small fossils. In fact chert is thought to have been produced by the shells of protozoa that contain silica as opposed to calcium carbonate. Because it is almost exclusively silica, it polishes very well. Chert is the light colored version of flint. Both were used by native Americans and the ancients to make arrowheads and spearpoints.

Examples of Chert

(Chert)

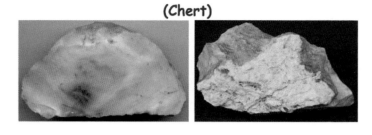

3. Flint - is a hard, sedimentary cryptocrystalline form of the mineral quartz, categorized as a variety of chert. It occurs chiefly as nodules and masses in sedimentary rocks, such as chalks and limestones. Inside the nodule, flint is usually dark grey, black, green, white, or brown in colour, and often has a glassy or waxy appearance. A thin layer on the outside of the nodules is usually different in color, typically white and rough in texture. From a petrological point of view, "flint" refers specifically to the form of chert which occurs in chalk or marly limestone. Similarly, "common chert" (sometimes referred to simply as "chert") occurs in limestone.

Examples of Flint

 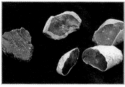 

(Flint)

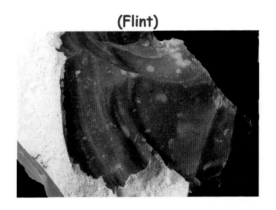

4. **Halite** – is rock salt, a sedimentary precipitate rock indicating a lot of super saturated salt water, even more than the current oceans. Halite is also a mineral, one of the few substances that shares both the distinction of being both a rock and a mineral.

Examples of Halite

5. **Gypsum** – is a sedimentary precipitate rock of super saturated calcium sulfate. Gypsum is abundant and comes in the a number of forms and colors. It is both a rock and a mineral.

Examples of Gypsum

6. Travertine – is a type of precipitate limestone made when acidic water passes through limestone and then deposits (precipitates) the residual in the form of porous layers often in banded iron oxide. Pure travertine is white. A strongly banded travertine is mistakenly called onyx. Rather it should be clarified to "travertine onyx". Travertine is most associated with the Mammoth Terraces in Yellowstone National Park.

Examples of Travertine

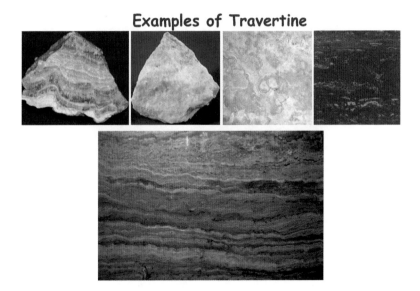

The Biochemical Sedimentary Rocks:
1. **Peat** – Peat is a mass of unconsolidated vegetable matter, which has accumulated under water, and in which the original plant remains are still, at least in part, discernable.

Examples of Peat

2. Lignite – a type of coal which is more compact than peat, and is found buried to some depth under layers of clay or sandstone. It is dark brown to black in color, and still retains pretty clear traces of the plants from which it was dervied. It is the lowest grade of coal used for industrial purposes.

Examples of Lignite Coal

3. Bituminous – a coal that is compact, black in color, and breaks readily, but does not crumble as easily as lignite. It is also the product of plant remains. It contains considerable water, and still has some hydrogen and oxygen compounds in it. It is believed that pressure has made it compact and nearly all traces of the original plants have disappeared.

Examples of Bituminous Coal

Anthracite (metamorphosed coal) – is the most desirable commercial grade coal, and is hard, lustrous and breaks with a concoidal fracture (like fractured glass).

Examples of Anthracite Coal

Special Rocks

Concretions

A concretion is a hard, compact mass of sedimentary rock formed by the precipitation of mineral cement within the
spaces between the sediment grains. Concretions are often ovoid or spherical in shape, although irregular shapes also occur. The word 'concretion' is derived from the Latin, *con* meaning 'together', and *crescere* meaning 'to grow'. Concretions are found within layers of sedimentary strata. There have been many ideas advanced about the origin of concretions, but no one definite thing stands out. Concretionary "cement" often makes the concretion harder and more resistant to weathering than the host stratum. There is an important distinction to draw between concretions and nodules. Concretions are formed from mineral precipitation around some kind of nucleus. A nodule is a small rock or mineral cluster.

Examples of Concretions

Concretions and nodules are fun to collect because there is so much variety and no two are exactly alike.

Nodules - clay, shale or mud filled with crystals or even fossils:

Examples of Nodules

 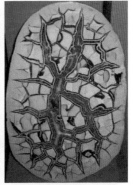

Examples of Nodules with Fossils

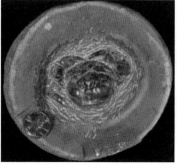

Geodes – the word means, "earthlike". They can be both sedimentary and volcanic. No one seems to have a definitive explanation for their origin. They are fun to collect and either break open to reveal crystals or to cut open with a saw and polished.

Examples of Geodes

Examples of Volcanic Geodes and Nodules

Volcanic geodes – crystal-lined volcanic nodules

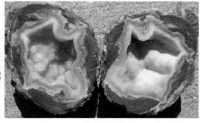

Volcanic nodules – solid, thin-shelled masses of agate

Volcanic "thunder eggs" or solid geodes – nodules or balls of rhyolite lava filled with one or more of the following: agate, opal, jasper

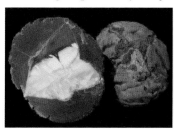
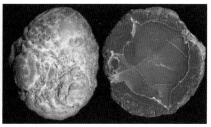

Large volcanic agate nodules from Brazil or Uruguay

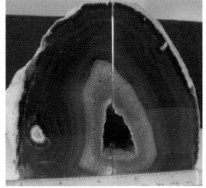
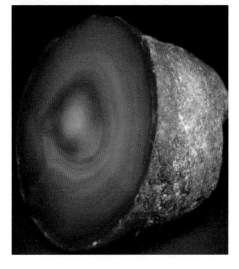

Collecting and building your rock collection

Rock collecting helps build your understanding of rocks and minerals. Getting out into the field will enable you, with time, to discover the varieties and similarities of the many rocks and minerals all around us. It will help you to discover the different environments all around us and will stimulate your thinking regarding the Genesis view of Creation and the Global Flood. Here are some practical tips for accomplishing this:

1. Location – always remember to carry a small notepad with you into the field so that you can keep track of the location where you found your treasures. You can always identify your find later, but you cannot always remember where you found it.

2. Notebook – carry a small notebook especially for making notes about your observations and thoughts. I don't know how many times I have thought of something in the field and did not write it down, and then promptly forgot it – and it was a great thought!

3. Display – you don't have to be fancy. Collect egg cartons. They make great storage containers. You can paint your egg cartons to match your rock type. I use a color code system for my rocks and minerals. A PowerPoint program will help you customize labels exactly the way you want them.

4. Roadguides – I use the Roadside Geology series. Although they are evolutionary, they do tell you where the rocks are located for each state.

5. Rockhounding areas – I have found that the best places to look for rocks are in road cutouts along US highways and at the bends of streams or rivers where rocks have been deposited by flowing rivers or streams.

6. Framework – always remember that the science of rocks and minerals does not change between Biblical Geologists and Secular Geologists. It is the worldview or the framework that is the issue. Master the first 11 chapters of Genesis, Romans chapter 1, 2 Peter chapter 3 and Psalm 104:5-9. These Scriptures will help you develop a solid worldview or framework for studying the origin of rocks and minerals.

Made in the USA
Charleston, SC
19 October 2015